A programmed text in statistics

BOOK 3
The t-test and χ^2 Goodness of fit

Other Books by G. B. Wetherill

Elementary Statistical Methods.
Sequential Methods in Statistics.
Sampling Inspection and Quality Control.

A programmed text
in statistics

BOOK 3
The t-test and χ^2 Goodness of fit

J. HINE
Formerly Research Officer
University of Bath

G.B. WETHERILL
Professor of Statistics
University of Kent, Canterbury

CHAPMAN AND HALL
LONDON

First published 1975
by Chapman and Hall Ltd.,
11 New Fetter Lane, London EC4P 4EE

Typeset by
E.W.C. Wilkins Ltd.
London & Northampton
Printed in Great Britain by
Whitstable Litho

ISBN 0 412 13740 2

© 1975 J. Hine and G. B. Wetherill

This paperback edition is sold subject to the condition that it shall not, by way of trade or otherwise, be lent, re-sold, hired out, or otherwise circulated without the publisher's prior consent in any form of binding or cover other than that in which it is published and without a similar condition including this condition being imposed on the subsequent purchaser.

All rights reserved. No part of this book may be reprinted, or reproduced or utilized in any form or by any electronic, mechanical or other means, now known or hereafter invented, including photocopying and recording, or in any information storage and retrieval system, without permission in writing from the Publisher.

Distributed in the U.S.A.
by Halsted Press, a Division
of John Wiley & Sons, Inc., New York

Library of Congress Catalog Number 75–1230

Contents

	page
Preface	vii
Section 1	1
Students t-distribution	1–12
Use of t-tables [1], *confidence intervals using t* [6],	

BOOK 3. THE t-TEST AND χ^2 GOODNESS OF FIT

Corrigenda

Page

1	Line 21	for $N(\mu, \sigma_{\bar{x}}^2) = \dfrac{S}{\sqrt{n}}$, where $\sigma_{\bar{x}} = \dfrac{S}{\sqrt{n}}$; read $N(\mu, \sigma_{\bar{x}}^2)$ where $\sigma_{\bar{x}} = \dfrac{S}{\sqrt{n}}$.
9	Line 4	for $\sigma_{\bar{x}} = s\sqrt{\dfrac{1}{n_1} + \dfrac{1}{n_2}}$; read $\sigma_{\bar{x}} = S\sqrt{\left(\dfrac{1}{n_1} + \dfrac{1}{n_2}\right)}$.
10	Frame 17	In the first response, *for* where $\sigma_{\bar{x}}\sqrt{\left(\dfrac{1}{n_1} + \dfrac{1}{n_2}\right)}$; read $\sigma_{\bar{x}} = S\sqrt{\left(\dfrac{1}{n_1} + \dfrac{1}{n_2}\right)}$.
11	Frame 19	*for* where $\sigma_{\bar{x}} = \sqrt{\left(\dfrac{1}{n_1} + \dfrac{1}{n_2}\right)}$; read $\sigma_{\bar{x}} = S\sqrt{\left(\dfrac{1}{n_1} + \dfrac{1}{n_2}\right)}$.
32	Line 16	*for* $\chi^2 = \dfrac{12}{6.5} + \dfrac{12}{7.5} + \dfrac{12}{7.5} + \dfrac{12}{8.5}$; *read* $\chi^2 = \dfrac{1^2}{6.5} + \dfrac{1^2}{7.5} + \dfrac{1^2}{7.5} + \dfrac{1^2}{8.5}$.

Exercises for Section 2   33

   Physical sciences and engineering   33

   Biological sciences   34

   Social sciences   35

Solutions to Exercises, Section 1   36

   Physical sciences and engineering   36

   Biological sciences   38

   Social sciences   39

Solutions to Exercises, Section 2   42

   Physical sciences and engineering   42

   Biological sciences   45

   Social sciences   47

Tables   50

   Squares and square roots   50

   Students t-distribution   51

   Reciprocals   52

   χ^2 distribution   53

Preface

This project started some years ago when the Nuffield Foundation kindly gave a grant for writing a programmed text to use with service courses in statistics. The work carried out by Mrs. Joan Hine and Professor G. B. Wetherill at Bath University, together with some other help from time to time by colleagues at Bath University and elsewhere. Testing was done at various colleges and universities, and some helpful comments were received, but we particularly mention King Edwards School, Bath, who provided some sixth formers as 'guinea pigs' for the first testing, the Bishop Lonsdale College of Education, and Bradford University.

Our objectives in the text are to take students to the use of standard t, F and χ^2 tests, with some introduction to regression, assuming no knowledge of statistics to start, and to do this in such a way that students attain some degree of understanding of the underlying reasoning.

Service courses are often something of a problem to statistics department. The classes are frequently large, and the students themselves are very varied in their background of mathematics. Usually, a totally inadequate amount of time is allocated for an extensive syllabus. The solution offered here is to use the programmed text in place of lectures, and have a weekly practical class at which students work through the exercises given at the end of each section. The available staff effort can then be placed in helping individuals out of particular difficulties, rather than in mass lecturing.

At one time we envisaged producing about three short video taped lectures to discuss some of the important concepts: random variation, probability distribution and sampling distribution. Unfortunately finance could not be obtained for this, and attempts to do it 'on the cheap' proved impossible, although it is to be hoped that this will be achieved in the future.

The main text units are written using examples drawn from many fields of application. In addition, several different sets of exercises are provided, each dealing with a particular field, so that students can apply the techniques within their own sphere of interest. The alternative sets of exercises can either be ignored, or used as 'spares'. Exercises should be written in a notebook, and submitted regularly. Further, students should make their own summaries in a notebook, using as a basis the 'summaries' given at the end of each section.

Thos using the text on their own should train themselves to complete all exercises relevant to their particular field of interest, and make notes as described above.

We have endeavoured to keep the mathematical level of the text low; however, it is inevitable that some concepts and notation have to be included. We would particularly mention the use of the exponential (e)

in Book 2 section 2. It would be impractical to describe this function in the text but any one not familiar with it can either consult their tutor or read about it in any basic mathematics book. Other notation is adequately described in the text.

The production of this text in four units is designed for several reasons. Firstly, it avoids giving the student a volume of such a size that it would be discouraging. Secondly, if some lecturers prefer not to use certain sections, they can easily replace these by standard lectures, or write their own text units. Thirdly, it facilitates use of the text for slightly different syllabuses. Books 3 and 4 can be used individually to study the use of t, F, and to give a simple introduction to regression. However, if this is done, the material presented in Books 1 and 2 must be covered in some other way.

It may be helpful here to explain something about the method of construction of the programmed texts. Once the objectives and entry behaviour are stated, the material to be covered is analysed and put into a 'flow chart'. Material not relevant is omitted. We have done all this for a basic common core of material, and clearly some students may require something more. In particular, some groups of students may require rather more on probability, and a coverage of statistical independence. We have prepared an alternative version of the earlier units, for those with greater mathematical ability, but this is not yet available commercially. Programmed texts on probability alone, however, are available from several sources.

Our thanks are due to Mrs. J. Honebon for her painstaking work in typing the manuscripts, and retyping them many times for the trials.

J.H.
G.B.W.
August 1974

HOW TO USE THE TEXT

Although there are prose passages, the bulk of the text is in programmed form. A statement is given on the left, and this includes gaps in which you have to judge the correct response. The column on the right gives the answer, and this should be covered until you have decided what response you are making.

NOTE: Students attempting this unit without having studied the first two volumes in the series must have covered elementary statistics up to the ideas of significance tests and confidence intervals for a normal mean, when σ is known.

Section 1

STUDENT'S t-DISTRIBUTION

Introduction

The sampling distribution of the means, $\bar{x}$, of samples of size n from an $N(\mu, \sigma^2)$ distribution is $N(\mu, \sigma_{\bar{x}}^2)$ where $\sigma_{\bar{x}} = \dfrac{\sigma}{\sqrt{n}}$.

We stated, in the section ESTIMATION, that if σ is unknown the sample standard deviation, s, is a good estimate of σ.

However, you may have noticed in the section TESTING HYPOTHESES that we never considered a problem in which σ was unknown. The reason for this is given below.

In order to transform an $N(\mu, \sigma^2)$ distribution to the standard normal we use the transformation $z = \dfrac{x - \mu}{\sigma}$.

However, if we use s as an estimate of σ, it can be shown that applying such a transformation to the sampling distribution of means does *not* produce the standard normal distribution. The distribution obtained in the STUDENT'S t-DISTRIBUTION. This distribution, unlike the Normal distribution, has an additional parameter, known as *the number of degrees of freedom* associated with it. You will remember that the formula for s^2 is

$$s^2 = \frac{1}{n-1} \sum (x - \bar{x})^2.$$

The quantity $n - 1$ which appears in the denominator is called the number of degrees of freedom and is usually denoted by the Greek letter ν (pronounced nu).

Summing up,

(i) the sampling distribution of means of samples of size n from an $N(\mu, \sigma^2)$ distribution where σ is unknown is

$$N(\mu, \sigma_{\bar{x}}^2) = \frac{s}{\sqrt{n}}, \text{ where } \sigma_{\bar{x}} = \frac{s}{\sqrt{n}}$$

(ii) **Transforming this distribution using the transformation**

$$\frac{\bar{x} - \mu}{\sigma_{\bar{x}}}$$

produces the STUDENT'S t-DISTRIBUTION with $n - 1$ DEGREES OF FREEDOM.

i.e. $t = \dfrac{\bar{x} - \mu}{\sigma_{\bar{x}}}$ where $\sigma_{\bar{x}} = \dfrac{s}{\sqrt{n}}$

For large ν, i.e. $\nu > 30$, the t-distribution approaches the Standard Normal Distribution.

Graphs of the t-distribution with varying degrees of freedom are shown together with the Standard Normal Distribution in Fig. 1.1.

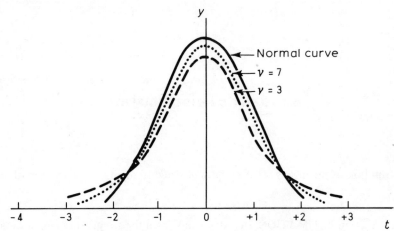

Fig. 1.1. Typical student's t-distribution

Use of t-tables

1.

In the t-distribution, as in the NORMAL DISTRIBUTION, probabilities are represented by the appropriate ... under the curve.	areas
However, it would be impractical to tabulate the areas in the same form as in the table of Standard Normal Curve Areas as this would necessitate a different table for each value of ... (THINK!).	ν (degrees of freedom)

2.

The table used in calculations involving the t-distribution is given on p. 51.

It can be seen that instead of tabulating values of an area corresponding to a particular t value, we tabulate values of ... corresponding to a particular	t area

3.

The sum, A, of the shaded areas represents the probability that a value lies (inside/outside) the range $-t \to +t$.
Consider the t-distribution with *3 d.f.* (degrees of freedom). From the t-tables, we see that, for this distribution, the probability of a value of t lying outside the limits -3.182 and $+3.182$ is *0.05*

	ν or d.f.	Probability ↓ A = 0.05	Probability ↓ A = 0.01
	1	12.706	63.657
	2	4.303	9.925
d.f.	3	3.182	5.841
	4	2.776	4.604
	5	2.571	4.032
	│	│	│

Similarly the probability of a value lying outside the range $-5.841 \to +5.841$ is ....

0.01

4.

For the *t-distribution with 10 d.f* the probability of a value lying outside the limits ... and ... is 0.05

-2.228 and $+2.228$

and
the probability of value lying outside the limits ... and ... is 0.01.

-3.169 and $+3.169$

5.

Alternatively, for the above distribution, the probability of a value lying *inside* the limits -2.228 and $+2.228$ is ...

$1 - 0.05 = 0.95$

and
the probability of a value lying *inside* the limits -3.169 and $+3.169$ is ...

0.99

Calculation of confidence intervals using *t*-distribution

6.

The values of t given in the table are, in fact, the values we use to calculate the 95 per cent and 99 per cent confidence intervals for μ
See Frame 7.

7.

Consider the following example:-

A sample of 16 variates has a mean of 34.86 and a standard deviation of 4.24. Find limits within which, 99 per cent of the time, we would be correct in saying that the population mean, μ, lies.

$n = ...;\quad \bar{X} = ...;\quad s = ...$	16; 34.86; 4.24
Since σ is unknown	
$$\sigma_{\bar{x}} = \frac{s}{\sqrt{n}}$$	
$$= \frac{...}{...}$$	$\dfrac{4.24}{4}$
$= ...$	$= 1.06$
and the transformation $\dfrac{34.86 - \mu}{1.06}$ $\left(\text{i.e. } \dfrac{\bar{X} - \mu}{\sigma_{\bar{x}}}\right)$	
produces the ...-distribution with	t
... d.f.	15
From the tables we see that for this distribution the probability of a value lying between ... and ... is 0.99.	-2.947 and $+2.947$
Substitution of these t values into the transformation $t = \dfrac{34.86 - \mu}{1.06}$ gives the 99 per cent confidence interval for μ	
Hence 99 per cent confidence interval for μ is ... $\rightarrow$ ...	$34.86 - 2.947 \times 1.06 \rightarrow$ $34.86 + 2.947 \times 1.06$

8.

In a sample of 9 variates the mean is 26.3 and the standard deviation 1.9

Calculate the 95 per cent confidence interval for the population mean μ.
Give full working.

If you have any difficulties with this question go to Frame 9. If not, go to Frame 11.	$26.3 - 0.63 \times 2.306 \rightarrow$ $26.3 + 0.63 \times 2.306$

9.

$n = ...;\quad \bar{X} = ...;\quad s = ...$ | 9; 26.3; 1.9

Since σ ... | σ is unknown

$$\sigma_{\bar{x}} = \frac{...}{...} \text{ (in symbols)}$$ | $\dfrac{s}{\sqrt{n}}$

$$= \frac{...}{...} \text{ (numerical values)}$$ | $\dfrac{1.9}{3}$

$$= ...$$ | 0.63

and the transformation $\dfrac{26.3 - \mu}{0.63}$ produces the ...-distribution with ... d.f. | t 8

From the tables we see that for this distribution the probability of a value lying between ... and ... is 0.95. | -2.306 and $+2.306$

Substitution of these values into the transformation
$t = \dfrac{26.3 - \mu}{0.63}$ gives the 95 per cent confidence interval for μ.

Therefore the 95 per cent confidence interval for μ is ... → ... | $26.3 - 0.63 \times 2.306$
 $\to 26.3 + 0.63 \times 2.306$

10.

Now calculate the 99 per cent confidence limits for μ using the data given in Frame 9.

Give full working.

| $26.3 - 0.63 \times 3.355$
 $\to 26.3 + 0.63 \times 3.355$

Testing hypotheses using t-distribution

Single sample problem

11.

We have already seen that for a t-distribution with 3 degrees of freedom, the probability of a value lying outside the limits -3.182 and $+3.182$ is 0.05 and the probability of a value lying outside the limits -5.841 and $+5.841$ is 0.01.

i.e. these limits are in fact the 5 per cent and 1 per cent ... levels for the given t-distribution. | significance

---

5

12.

Consider the following example:-

A production process gives components whose strengths are normally distributed with mean 40 lbs.
9 components are selected at random, the mean of this sample being 40.9 lb. and the standard deviation 1.2 lb.
Is there a significant difference between the sample and population means?

Firstly we must set up the ... i.e. assume that there (is a/is no) difference between the sample and population means.	null hypothesis is no
Next we have to find the ...	significance level

In this example,

$\mu = ...$	$\bar{X} = ...$	40	40.9
$s = ...$	$n = ...$	1.20	9

Since ... σ is unknown

$$\sigma_{\bar{x}} = \frac{...}{...} \text{ (in symbols)} \qquad \frac{s}{\sqrt{n}}$$

$$= \frac{...}{...} \text{ numerical values} \qquad \frac{1.20}{3}$$

$$= ... \qquad 0.40$$

and we transform to the ...-distribution t
with ... d.f.   8
For this example, the transformation is

$$t = \frac{\bar{X} - ...}{...} \text{ (insert numerical values)} \qquad t = \frac{\bar{X} - 40}{0.40}$$

Hence when $\bar{X} = 40.9$, $t = ...$

$$t = \frac{40.9 - 40}{0.40}$$

$$= 2.25$$

From the table we see that for the t-distribution with 8 d.f. the 1 per cent significance level is $t = -3.355$ and $t = +3.355$ and the 5 per cent significance level is
$t = ...$ and $t = ...$ $t = -2.306$ and
                            $t = +2.306$

Therefore, the value $t = 2.25$ lies ....   inside the 5 per cent level.

Therefore the difference between $\bar{X}$ and μ is not significant at the ... level.   5 per cent
i.e. there is ... evidence of a real difference between $\bar{X}$ and μ no
and so we ... the null hypothesis.   accept

13.

Thus, if σ is unknown, in order to answer the question 'Could this sample have come form a given population?' we have to work the following steps:-

STEP 1. Set up ... ... i.e. assume that	null hypothesis the sample was drawn from the given population i.e. there is no real difference between the sample and population means.
STEP 2. Transform to ...-distribution with ... d.f. using the transformation ... ... where $\sigma_{\bar{x}}$ = ... and hence, using *t*-tables, find	t $n-1$ $t = \dfrac{\bar{X} - \mu}{\sigma_{\bar{x}}}$ where $\sigma_{\bar{x}} = \dfrac{s}{\sqrt{n}}$ significance level
STEP 3. Draw conclusions. If difference between $\bar{X}$ and μ is *significant* at (i) 1 per cent level – ... ... evidence of real difference. Therefore ... null hypothesis. (ii) 5 per cent level – ... evidence of real difference. Therefore ... null hypothesis. If difference is *not significant* at 5 per cent level – ... evidence of a real difference. Therefore ... null hypothesis.	 almost conclusive reject reasonable reject no accept.

14.

A coin-operated, soft-drink machine was designed to discharge, on the average, 7 ounces of beverage/cup. To test the machine, 9 cupsful of beverage were drawn from the machine and measured. The mean and standard deviation of these measurements were 6.8 ounces and 0.12 ounces respectively. Do these data present sufficient evidence to indicate that the mean discharge differs from 7 ounces?

                 Give full working.

If you have any difficulties with this question go to Frame 15. If not, go to Frame 16.	Almost conclusive evidence of real difference between $\bar{X}$ and μ.

15.

Step 1. Set up ... i.e. assume that the sample mean (does/does not) differ from 7 ounces.	null hypothesis does not

Step 2. Transform to ...-distribution
with ... d.f. using the transformation

... ... where $\sigma_{\bar{x}} = ...$
  (in symbols).

In this example,
$\bar{X} = ...$; $\mu = ...$;
$s = ...$; $n = ...$; $\sigma_{\bar{x}} = ...$
Therefore $t = \dfrac{... - ...}{...} = ...$

From the *t*-tables we see that for the *t*-distribution
with 8 d.f. 1 per cent significance level is
$t = ...$ and $t = ...$
5 per cent significance level is
$t = ...$ and $t = ...$

The value $t = -5.0$ lies ....

Therefore the difference between $\bar{X}$ and μ is
significant at ... level
i.e. there is ... evidence of real difference between $\bar{X}$ and μ.
Therefore we ... the null hypothesis.

---

t
8

$t = \dfrac{\bar{X} - \mu}{\sigma_{\bar{x}}}$

where $\sigma_{\bar{x}} = \dfrac{s}{\sqrt{n}}$

6.8   7.0
0.12 9 0.04
$t = \dfrac{6.8 - 7.0}{0.04}$

$= -5.0$

-3.355 and $+3.355$

-2.306 and $+2.306$

outside the 1 per cent level.

1 per cent
almost conclusive
reject

16.

A manufacturer of light bulbs claims that his bulbs will burn on the average 500 hours. To maintain this average he tests 25 bulbs each month. What conclusions should he draw from a sample that has a mean $\bar{X} = 518$ hours and a standard deviation $s = 40$ hours.
  Give full working.

A complete solution is given on p.11.

Two sample problems

INTRODUCTION

Suppose we have two samples

	size:	n_1	n_2
	mean:	$\bar{X}_1$	$\bar{X}_2$
	standard deviation:	s_1	s_2

and we wish to test the hypothesis that these samples were drawn from the same population (or from populations with the same mean).
  We have learnt that the sampling distribution of the difference between means of samples of size n_1 and n_2 is $N(0, \sigma_{\bar{x}}^2)$ where $\sigma_{\bar{x}} = \sigma \sqrt{\left(\dfrac{1}{n_1} + \dfrac{1}{n_2}\right)}$.

However, if **σ** is unknown, we have to use an estimate. The best estimate is given by the following equation:-

$$s = \sqrt{\frac{[(n_1-1)s_1^2 + (n_2-1)s_2^2]}{n_1+n_2-2}}.$$

Transforming using the transformation $t = \dfrac{(\overline{X}_1 - \overline{X}_2) - 0}{\sigma_{\bar{x}}}$

where $\sigma_{\bar{x}} = s\sqrt{\dfrac{1}{n_1} + \dfrac{1}{n_2}}$

and $s = \sqrt{\dfrac{[(n_1-1)s_1^2 + (n_2-1)s_2^2]}{n_1+n_2-2}}$

gives the *t*-distribution with $n_1 + n_2 - 2$ **DEGREES OF FREEDOM**.

Hence, using the *t*-tables we can find level of significance of difference between $\overline{X}_1$ and $\overline{X}_2$ and make the appropriate deductions concerning the null hypothesis.

17.

Is there any evidence to assume that the samples given were drawn from the same population?

	Sample 1	Sample 2
size:	4	10
mean:	131	135
standard deviation:	2.3	1.9

Null hypothesis: Assume .... the samples were drawn from the same population.

In this example –

$n_1 = ...;\quad \overline{X}_1 = ...;\quad s_1 = ...$ 4   131   2.3
$n_2 = ...;\quad \overline{X}_2 = ...;\quad s_2 = ...$ 10   135   1.9

The best estimate of σ is given by

$$s = \sqrt{\frac{[(n_1-1)s_1^2 + (n_2-1)s_2^2]}{n_1+n_2-2}}$$

Substituting numerical values gives

$s = ...$ $\sqrt{\dfrac{[(3 \times 5.29) + (9 \times 3.61)]}{12}}$

$= ...$ $\sqrt{\dfrac{48.36}{12}} = \sqrt{4.03}$

$\simeq ....$ $\simeq 2$

We can now transform to the ...-distribution with ... d.f. t;   12 i.e. $(n_1 + n_2 - 2)$

The transformation is

$$t = \frac{\cdots}{\cdots} \text{ where } \sigma_{\bar{x}} = \cdots$$

(in symbols).

	$t = \dfrac{(\bar{X}_1 - \bar{X}_2) - 0}{\sigma_{\bar{x}}}$ where $\sigma_{\bar{x}} \sqrt{\left(\dfrac{1}{n_1} + \dfrac{1}{n_2}\right)}$

Therefore, for the data given,

$$t = \frac{\cdots}{\cdots} \text{ (numerical values)}$$

$$t = -3.39.$$

	$t = \dfrac{(131 - 135) - 0}{2\sqrt{\left(\dfrac{1}{4} + \dfrac{1}{10}\right)}}$

From the t-tables we see that for the t-distribution with 12 d.f.
1 per cent significance level is $t = \ldots$ and $t = \ldots$ -3.055 and $+3.055$
5 per cent significance level is $t = \ldots$ and $t = \ldots$ -2.179 and $+2.179$
The value $t = -3.39$ lies … …   outside the 1 per cent level
Therefore the difference between $\bar{X}_1$ and $\bar{X}_2$ is significant
at … level.   1 per cent

Hence, what conclusions can we draw?   There is almost conclusive evidence of a real difference between the means ∴ reject the null hypothesis.

18.

In a random sample of 10 frogs in the natural state, the average life span was 13 years with a standard deviation of 3 years, while in a random sample of 20 well-cared-for frogs, the average life span was 15 years with standard deviation 4 years. Is one justified in concluding that the mean life of well-cared-for frogs is different from that of frogs in the natural state?

Give full working.

	No evidence to assume that the mean life of well-cared-for frogs is different from that in the natural state.
If you have any difficulties go to Frame 19. If not, go to Frame 20.	

19.

Null hypothesis: ….   Assume that the samples were drawn from populations with equal means.

In this example
$n_1 = \ldots; \quad \bar{X}_1 = \ldots; \quad s_1 = \ldots$ 10   13   3
$n_2 = \ldots; \quad \bar{X}_2 = \ldots; \quad s_2 = \ldots$ 20   15   4

The best estimate of σ is given by $s = \ldots$ (in symbols).	$s = \sqrt{\dfrac{[(n_1-1)s_1^2 + (n_2-1)s_2^2]}{n_1 + n_2 - 2}}$
Therefore, in this example	
$s = \ldots$	$s = \sqrt{\dfrac{9 \times 9 + 19 \times 16}{28}}$
$= \ldots$	$\sqrt{13.75}$
$s = 3.7$	
We can now transform to the t-distribution with $\ldots$ d.f.	28
The transformation is $t = \dfrac{\ldots}{\ldots}$	$t = \dfrac{(\overline{X}_1 - \overline{X}_2) - 0}{\sigma_{\bar{x}}}$
where $\sigma_{\bar{x}} = \ldots$ (in symbols).	where $\sigma_{\bar{x}} = \sqrt{\left(\dfrac{1}{n_1} + \dfrac{1}{n_2}\right)}$
Therefore, for this problem	
$t = \dfrac{\ldots}{\ldots}$	$\dfrac{(13 - 15) - 0}{3.7\sqrt{\left(\dfrac{1}{10} + \dfrac{1}{20}\right)}}$
$= -1.40$ (to 2 d. pl.).	
For the t-distribution with 28 d.f. 1 per cent significance level is $t = \ldots$ and $t = \ldots$	-2.763 and $+2.763$
5 per cent significance level is $t = \ldots$ and $t = \ldots$	-2.048 and $+2.048$
The value $t = -1.40$ lies $\ldots$.	within 5 per cent level
Therefore, the difference is $\ldots \ldots$ at 5 per cent level.	not significant
i.e. $\ldots$ evidence of real difference between $\overline{X}_1$ and $\overline{X}_2$ and we $\ldots$ null hypothesis.	no accept

20.

A University investigation, conducted to determine whether car ownership was detrimental to academic achievement, was based upon two random samples of 15 male students, each drawn from the student body. The grade-point average for the $n_1 = 15$ non-car-owners possessed an average and variance equal to $\overline{X}_1 = 2.70$ and $s_1^2 = 0.36$ as opposed to a $\overline{X}_2 = 2.54$ and $s_2^2 = 0.40$ for the $n_2 = 15$ car-owners. Do the data present sufficient evidence to indicate a difference in the mean achievement between car-owners and non-car-owners?

Give full working.

A complete solution is given on p. 12.

Solution to problem given in Frame 16

Null hypothesis: Assume that sample mean does not differ from 500 hours.

$$\overline{X} = 518 \quad \mu = 500$$

$$s = 40 \quad n = 25 \text{ therefore } \sigma_{\bar{x}} = \dfrac{40}{5} = 8$$

Transform to *t*-distribution with 24 d.f. using transformation

$$t = \frac{\overline{X} - 500}{8}$$

Therefore, when $\overline{X} = 518$, $t = \dfrac{518 - 500}{8}$

$$= 2.25$$

For the *t*-distribution with 24 d.f.
1 per cent significance level is $t = -2.797$ and $t = +2.797$
5 per cent significance level is $t = -2.064$ and $t = +2.064$.

The value $t = 2.25$ lies within 1 per cent level but outside 5 per cent level therefore difference between $\overline{X}$ and μ is significant at 5 per cent level. i.e. there is reasonable evidence of a real difference between $\overline{X}$ and μ, therefore we reject the null hypothesis.

Manufacturer is likely to conclude that his bulbs are a better product than he thought!

Solution to problem given in Frame 20

Null hypothesis: Assume there is no difference between the mean achievements of car-owners and non-car owners.

$$n_1 = 15, \quad \overline{X}_2 = 2.70, \quad s_1^2 = 0.36$$
$$n_2 = 15, \quad \overline{X}_2 = 2.54, \quad s_1^2 = 0.40$$

Best estimate of σ is given by

$$s = \sqrt{\left(\frac{(14 \times 0.36) + (14 \times 0.40)}{28}\right)}$$

$$= \sqrt{0.38}$$

$$= 0.62.$$

Now transform to *t*-distribution with 28 d.f.

$$t = \frac{(2.70 - 2.54) - 0}{0.62 \sqrt{\left(\dfrac{1}{15} + \dfrac{1}{15}\right)}}$$

$$= 0.70.$$

For *t*-distribution with 28 d.f.
1 per cent significance level is $t = -2.763$ and $t = +2.763$
5 per cent significance level is $t = -2.048$ and $t = +2.048$.

The value $t = 0.7$ lies within the 5 per cent level.
Therefore, difference is not significant at 5 per cent level.
i.e. no evidence of real difference between $\overline{X}_1$ and $\overline{X}_2$ and we accept the null hypothesis.

Summary

NOTE: **The *t*-distribution is only used if σ is unknown.**

1. Single sample problem

Population: mean μ
Sample: mean $\overline{X}$, size n
standard deviation s

Could this sample have been drawn from the given population?

Null hypothesis: Assume that sample was drawn from given population.

Level of significance:

Transform to *t*-distribution with $(n-1)$ d.f. using the transformation

$$t = \frac{\overline{X} - \mu}{\sigma_{\bar{x}}} \quad \text{where} \quad \sigma_{\bar{x}} = \frac{s}{\sqrt{n}}$$

and find level of significance using *t*-tables.

Conclusion:

Significance at 1% level — almost conclusive evidence of real difference between $\overline{X}$ and μ.
Therefore reject null hypothesis.
Significance at 5% level — reasonable evidence of real difference between $\overline{X}$ and μ.
Therefore reject null hypothesis.
No significance — No evidence of real difference between $\overline{X}$ and μ.
Therefore accept null hypothesis.

2. Two sample problem

Samples: mean $\overline{X}_1$, standard deviation s_1, size n_1
mean $\overline{X}_2$, standard deviation s_2, size n_2

Could these samples have been drawn from the same population?
(or from populations with same mean).

Null hypothesis: Assume samples were drawn from same population.

Level of significance:

Transform to *t*-distribution with $(n_1 + n_2 - 2)$ d.f. using the transformation

$$t = \frac{(\overline{X}_1 - \overline{X}_2) - 0}{\sigma_{\bar{x}}}$$

$$\text{where} \quad \sigma_{\bar{x}} = s\sqrt{\left(\frac{1}{n_1} + \frac{1}{n_2}\right)}$$

$$\text{and} \quad s = \sqrt{\frac{[(n_1-1)s_1^2] + [(n_2-1)s_2^2]}{n_1 + n_2 - 2}}$$

and find level of significance using *t*-tables.

Conclusion

Significance at 1% level — almost conclusive evidence of real difference between $\bar{X}_1$ and $\bar{X}_2$.
Therefore reject null hypothesis.
Significance at 5% level — reasonable evidence of real difference between $\bar{X}_1$ and $\bar{X}_2$.
Therefore reject null hypothesis.
No significance — no evidence of real difference between $\bar{X}_1$ and $\bar{X}_2$.
Therefore accept null hypothesis.

Exercises for Section 1

PHYSICAL SCIENCES AND ENGINEERING

Student's *t*-distribution

1. The percentage of copper in a certain chemical is measured 5 times. The sample mean and standard deviation are found to be 14.1 and 2.1 respectively. Give a 95 per cent confidence interval for the true percentage of copper.

2. A new method for making concrete has been proposed. To test whether the new method has increased the compressive strength, five sample blocks are made by each method. The following results were obtained.

	Mean	Standard deviation (in lbf/in^2)
New method	143	8.28
Old method	138	11.58

   Would you say that the new method has increased the compressive strength?

3. A sample of 9 measurements of the percentage of manganese in ferromanganese has a mean of 84.0 and a standard deviation of 1.2. Test the null hypothesis that the true percentage of manganese is 80.0.

BIOLOGICAL SCIENCES

Student's *t*-distribution

1. The percentage of copper in a certain chemical is measured 4 times. The sample mean and standard deviation are found to be 14.1 and 2.1 respectively. Give a 95 per cent confidence interval for the true percentage of copper.

2. The following results were obtained in a comparison of two supposedly soporific drugs A and B. Each patient received each drug once, the order being randomized, and the number of hours additional sleep recorded.
Compare the drugs.

	No. of patients	Mean	Standard deviation
Drug A	10	0.75	1.70
Drug B	10	2.33	2.10

3. Many sea animals of a certain species found off the coast of Scotland had a mean body length of 10.1 mm. Specimens were then found in the water in another area and the lengths of 9 of these recorded, giving the following results.

Sample mean 11.5 mm; sample standard deviation 1.4 mm. Is the mean body length in this area different from that in Scotland?

SOCIAL SCIENCES

Student's *t*-distribution

1. A sample of four grapefruit has a mean weight of 15.00 ounces and a standard deviation of 1.4 ounces. Calculate the 95 per cent confidence interval for the mean weight of all grapefruit in the shipment.

2. A comparison of reaction times for two different stimuli in a psychological word-association experiment produced the following results:

	No. in sample	Mean	Standard deviation
Stimulus 1	8	1.9	0.8
Stimulus 2	8	2.6	0.9

Do the data present sufficient evidence to indicate a difference in mean reaction time for the two stimuli?

3. A very costly experiment has been conducted to evaluate a new process for producing synthetic diamonds. 5 diamonds have been generated by the new process; the mean and standard deviation of this sample being 0.62 and 0.06 carat respectively.

A study of the process cost indicates that the average weight of the diamonds must be greater than 0.5 carat in order that the process be operated at a profitable level. Do the five diamond weight measurements present sufficient evidence to indicate that the average weight of the diamonds produced by the process is significantly different from 0.5 carat?

Section 2

GOODNESS OF FIT

Introduction

We have in previous sections described methods of testing hypotheses concerning the differences between *single* values — for example, between a sample mean and a hypothesis value.

In this section we shall be concerned with testing the difference between *sets* of values. The following example illustrates the type of problem in which we shall be interested.

Example:
A dice is tossed 120 times; the results are shown in the table below.

Outcome	1	2	3	4	5	6
Obs. freq.	20	22	17	18	19	24

Do these results indicate that the dice is biased ?

1.

In order to answer this question we compare the set of *observed* frequencies with the set of *expected* frequencies which we can calculate. If the dice is unbiased we (would/would not) expect each face to occur an equal number of times.

Thus, in 120 tosses we would expect each face to occur ... (how many?) times.

would

20   i.e. $\dfrac{120}{6}$

2.

We can now construct a table showing the observed (*O*) and expected (*E*) frequencies.
Complete the table below.

Outcome	1	2	3	4	5	6	total
Obs. freq. (*O*)	20	22	17	18	19	24	120
Ex-. freq. (*E*)	..	..	..	..	..	..	120

20 20 20 20 20 20

3.

We next have to test the *goodness of fit* between the above two sets of frequencies. i.e. we test the null hypothesis that there (is/is no) difference between the two sets of values.

is no

4.

This test makes use of the X^2 (Chi-square) distribution with *n*-1 degrees of freedom where *n* is the number of outcomes.

For the dice example, we test using the ... distribution with ... (how many?) d.f.

χ^2 5 (since $n = 6$)

5.

In order to calculate a value of X^2 for a given set of values we first evaluate the expression

$$\frac{(O-E)^2}{E}$$

for each outcome and then sum over all outcomes.

$$\text{i.e.} \quad X^2 = \Sigma \frac{(O-E)^2}{E}$$

In the dice example,

$$\chi^2 = \frac{(20-20)^2}{20} + \frac{(22-20)^2}{20} + \frac{(17-20)^2}{20} + \ldots$$

$$+ \ldots + \ldots$$

$$= \frac{0 + 4 + \ldots + \ldots + \ldots + 16}{20}$$

$$= 1.7 \;.$$

$\frac{(18-20)^2}{20} +$

$\frac{(19-20)^2}{20} + \frac{(24-20)^2}{20}$

$9 + 4 + 1$

6.

In the table of χ^2 values (p. 53) as in the t-tables, the 1 per cent and 5 per cent levels of significance are tabulated for varying d.f.
For the χ^2 distribution with 5 d.f.
  1 per cent significance level is $\chi^2 = 15.09$
  5 per cent significance level is $\chi^2 = \ldots$. | 11.07

7.

The value $\chi^2 = 1.7$ (is/is not) significant at 5 per cent level. | is not, since it lies within 5% limit.

Therefore there (is/is no) evidence of a real difference between the two sets of values and we (accept/reject) the null hypothesis. | is no
| accept
i.e. we conclude that there (is/is no) evidence that the dice is biased. | is no

---

8.

The number of books borrowed from a public library during a particular week is given in the table below.
Test the hypothesis that the number of books borrowed does not depend on the day of the week.

	Mon.	Tue.	Wed.	Thu.	Fri.
No. of books borrowed.	135	108	120	114	146

Total 623

Set up null hypothesis:
i.e. assume number of books borrowed (does/does not) depend on the day of the week. | does not

Calculate expected frequencies.
Since we have assumed that the number of books borrowed is not dependent on the day of the week, we (would/would not) expect the number of books borrowed on each day of the week to be equal. | would

i.e. expected frequencies $= \dfrac{\ldots}{\ldots}$ | $\dfrac{623}{5}$

$= \ldots$. | 124.6

Next we construct the table of observed and expected frequencies.

Obs.	135	108	120	114	146
Exp.	124.6	124.6	124.6	124.6	124.6

and calculate χ^2

$$\chi^2 = \frac{(135-124.6)^2}{124.6} + \dots + \dots$$

$$+ \dots + \dots$$

$$= \frac{(10.4)^2 + \dots + \dots + \dots + \dots}{\dots}$$

$$= \frac{975.20}{124.6} = 7.8.$$

	$\frac{(108-124.6)^2}{124.6} + \frac{(120-124.6)^2}{124.6}$
	$+ \frac{(114-124.6)^2}{124.6} + \frac{(146-124.6)^2}{124.6}$
	$\frac{(16.6)^2 + (4.6)^2 + (10.6)^2 + (21.4)^2}{124.6}$

We now find level of significance using ... distribution with ... d.f $\qquad$ χ^2 4 (since $n = 5$)
From tables,
  1 per cent significance level $\chi^2 = 13.28$
  5 per cent significance level $\chi^2 = \dots$ $\qquad$ 9.49
The value $\chi^2 = 7.8$ (is/is not) significant at 5 per cent level. $\qquad$ is not since it lies within the limit $\chi^2 = 9.49$

Therefore, there is ... evidence of a real difference between the $\qquad$ no
observed and expected frequencies and we ... the null hypothesis. $\qquad$ accept
i.e. we conclude that there is ... evidence that the number of $\qquad$ no
books borrowed ... . $\qquad$ depends on the day of the week.

9.

From the two examples it can be seen that testing for goodness of fit using the χ^2 distribution involves the following steps.

STEP 1. Set up ... ... $\qquad$ null hypothesis

STEP 2. Calculate ... frequencies $\qquad$ expected

STEP 3. Calculate a χ^2 value using

$$\chi^2 = \sum \frac{\dots}{\dots} ..$$
$\qquad$ $\sum \frac{(O-E)^2}{E}$

STEP 4. Determine level of significance using ... distribution $\qquad$ χ^2
with ... d.f. (where $n =$ number of outcomes) and $\qquad$ $n - 1$
hence draw conclusions.

10

A manufacturer of buttons wished to determine whether the number of defective buttons produced by three machines varied from machine to machine. The machines produce buttons at the same rate, and the number of defectives produced in a shift was counted. Results are given below.

Machine no.	1	2	3
No. of defectives.	16	24	8

Do these data present sufficient evidence to indicate that the number of defective buttons varies from machine to machine ?

Give full working.

	Significant at 5% level. Conclude that there is reasonable evidence that the number of defective buttons does vary from machine to machine.

If you had any difficulties with this problem go to Frame 11.
If not, go to Frame 13.

11.

STEP 1.   Set up null hypothesis.
i.e. assume that the number of defective buttons (does/does not) vary from machine to machine.

does not

STEP 2.   Calculate expected frequencies.
Since we have assumed that the number of defective buttons does not vary from machine to machine, expected frequencies

$$= \frac{\cdots}{\cdots} = \cdots \text{ (numerical values)}$$

$$\frac{48}{3} = 16$$

STEP 3.   Complete the following table

Obs.	16	...	...
Exp.	16	...	...

24   8
16   16

Therefore $\chi^2 = \frac{(16-16)^2}{16} + \cdots + \cdots$

$$\frac{(24-16)^2}{16} + \frac{(8-16)^2}{16}$$

$$= \frac{(\ldots)^2 + (\ldots)^2 + (\ldots)^2}{\ldots}$$

$$\frac{0 + (8)^2 + (8)^2}{16}$$

which gives
$\chi^2 = 8$.

STEP 4.   Test, using ... distribution
with ... d.f.
1 per cent significance level $\chi^2 = \ldots$
5 per cent significance level $\chi^2 = \ldots$
Thus the value $\chi^2 = 8$ is significant at ... level.
Therefore there is ... evidence of a real difference between ... and ... ...
and we ... the null hypothesis.
Therefore, we conclude ... ...

χ^2
2
9.21
5.99
5%
reasonable
observed
expected values
reject
that there is reasonable evidence that the number of defective buttons does vary from machine to machine.

12.

State the four steps in testing for goodness of fit using χ^2 distribution.

STEP 1.
STEP 2.
STEP 3.
STEP 4.

Refer back to Frame 9 for correct response.

13.

The following table shows the number of accidents/shift in a factory.

	Shift		
	Day	Evening	Night
Number of accidents.	1	7	7

Is there any evidence that the number of accidents is dependent upon shift?

Give full working

Complete solution given on p. 23.

Summary

Testing for goodness of fit using χ^2 distribution.

STEP 1.   Set up null hypothesis.

STEP 2.   Calculate expected frequencies.

STEP 3.   Calculate a χ^2 value using

$$\chi^2 = \sum \frac{(O-E)^2}{E}.$$

STEP 4.   Determine level of significance using χ^2 distribution with $n-1$ degrees of freedom (n being number of outcomes) and hence draw conclusions.

    Significance at 1 per cent level — almost conclusive evidence of real difference between observed and expected values.
        ∴ Reject null hypothesis.

    Significance at 5 per cent level — reasonable evidence of real difference between observed and expected values.
        ∴ Reject null hypothesis.

    No significance — no evidence of real difference between observed and expected values.
        ∴ Accept null hypothesis.

Solution to problem given in Frame 13

Null hypothesis: Assume that number of accidents is not dependent upon shift.

Expected frequencies: On the basis of the null hypothesis expected frequencies

$$= \frac{15}{3} = 5$$

Obs.	1	7	7
Exp.	5	5	5

$$\chi^2 = \frac{(1-5)^2}{5} + \frac{(7-5)^2}{5} + \frac{(7-5)^2}{5}$$

$$= \frac{16 + 4 + 4}{5}$$

$$= 4.8$$

Determine level of significance using χ^2 distribution with 2 d.f

1 per cent significance level $\chi^2 = 9.21$
5 per cent significance level $\chi^2 = 5.99$

The value $\chi^2 = 4.8$ is not significant at 5 per cent level.

Therefore, there is no evidence of a real difference between observed and expected values and we accept the null hypothesis.
i.e. we conclude that there is no evidence that the number of accidents is dependent upon shift.

CONTINGENCY TABLES

Introduction

In the examples we have considered so far, the frequencies have only been classified according to *one* criterion — number of books borrowed/day, number of defective buttons/machine, number of accidents/shift.
In the following work we shall consider the use of the χ^2 test in situations in which the frequencies are classified according to *two* criteria. An example of such a situation is given in Table 2.1.

Table 2.1.
Two groups of people were chosen at random, one from the east coast and one from the west coast of the United States and each person was classified as Protestant, Catholic or Jewish.

	Protestant	*Catholic*	*Jewish*	*Total*
East coast	182	215	203	600
West coast	154	136	110	400
Total	336	351	313	1000

In this case the two criteria are geographical location and religious affiliation.

1.

A table such as Table 2.1 is called a CONTINGENCY TABLE.

A contingency table having *r* rows and *c* columns is known as an $r \times c$ (read *r* by *c*) contingency table.

Thus contingency Table 2.1 is a ... × ... contingency table.

2×3
↑   ↑
no. of   no. of
rows.   columns.

2.

The χ^2 test is used to test the hypothesis that the two criteria are independent.
The method of carrying out this test is exactly similar to that for testing for goodness of fit.
            See Frame 3.

3.

The method for calculating expected frequencies in a contingency table is based upon the fact that we make the null hypothesis that the two criteria are independent.
For example, in Table 2.1, we assume that ... and ... are independent.

geographical location and religious affiliation

Hence considering first the east coast people we (would/would not) expect the following equation to hold —

would

$$\frac{\text{number of east coast Protestants}}{\text{total number of Protestants}}$$
$$= \frac{\text{number of east coast people}}{\text{total number of people}}$$

4.

From Table 2.1,
total number of Protestants = ...   336
number of east coast people = ...   600
total number of people   = ...   1000
Thus, if p = *expected* number of Protestants on east coast.

then $\frac{p}{...} = \frac{...}{...}$ (insert numerical values) $\frac{p}{336} = \frac{600}{1000}$

and so $p = 202$.

24

5.

Similarly, assuming independence of the two criteria, we would expect

$$\frac{\text{number of east coast Catholics}}{\text{total no. of people.}} = \frac{\ldots\ldots\ldots}{\ldots\ldots\ldots}$$

$\dfrac{\text{number of east coast Catholics}}{\text{total no. of Catholics}}$
$\dfrac{\text{number of east coast people}}{\text{total no. of people}}$

Thus if c = *expected* number of Catholics on east coast

then $\dfrac{c}{\ldots} = \dfrac{\ldots}{\ldots}$ (insert numerical values)

$\dfrac{c}{351} = \dfrac{600}{1000}$

and so $c = 211$.

6

From these results we can deduce a general formula for calculating the expected frequencies in a contingency table:-

If R = row total, C = column total and T = grand total,

then expected frequency = $\dfrac{\ldots \times \ldots}{\ldots}$

$\dfrac{R \times C}{T}$

7

Use the above formula to calculate the expected number of Protestant people on the west coast.

exp. freq. = $\dfrac{336 \times 400}{1000}$

= 134

8.

The remaining expected frequencies can be found by subtraction. For example, by subtracting the sum of the expected frequencies of Protestants and Catholics on the east coast from the total number of east coast people, we obtain the expected number of ... .

Jewish people on the east coast.

9.

Therefore expected frequency of Jewish people on the east coast
= ... − ... − ...
= ... .

$600 - 202 - 211$
= 187

10.

Thus, although we could compute the above expected frequency using the formula

$$\text{expected frequency} = \frac{R \times C}{T}$$

the simpler method is by subtraction.

Having calculated the expected frequencies for the east coast we (can/cannot) calculate the remaining expected frequencies for the west coast by subtraction.

can, since expected frequency for west coast = column frequency − expected frequency for east coast.

(THINK!)

11.

In the table below we have recorded the observed frequencies together with the expected frequencies (in brackets).
Fill in any missing values.

Observed and Expected Frequencies.

	Protestant	Catholic	Jewish	Total
E. coast	182(202)	215(211)	203(187)	600
W. coast	154(134)	136(...)	110(...)	400
Total	336	351	313	1000

(140)   (126)
i.e. ↑   i.e. ↑
351−211   313−187

12

We are now in a position to calculate a value of χ^2.

As in the test for goodness of fit

$$\chi^2 = \sum \frac{...}{...}$$

$$\frac{(O-E)^2}{E}$$

Therefore in our example,

$$\chi^2 = \frac{(182-202)^2}{202} + \frac{(215-211)^2}{211}$$

$$... + ... +$$

$$\frac{(203-187)^2}{187} + \frac{(154-134)^2}{134}$$

$$... + ...$$

$$+ \frac{(136-140)^2}{140} + \frac{(110-126)^2}{126}$$

which gives $\chi^2 = 8.56$ (to 2 d. pl.).

13

In testing for goodness of fit we found the level of significance using the χ^2 distribution with (...) d.f. where n = number of outcomes. | $n - 1$

However when applying the χ^2 test to an $r \times c$ contingency table, we find the level of significance using the χ^2 distribution with $(r - 1) \times (c - 1)$ d.f.

Therefore in our example we have to find the significance level using the χ^2 distribution with ... (how many?) d.f.

$$1 \times 2 = 2$$
$$\uparrow \qquad \uparrow$$
$$r-1 \qquad c-1$$

14.

Thus,
    1 per cent significance level $\chi^2 = 9.21$
    5 per cent significance level $\chi^2 = \ldots$.   5.99

Therefore value $\chi^2 = 8.56$ is significant at ... level .   5%

Therefore there is ... evidence of a real difference between observed and expected values and we ... the null hypothesis.   reasonable / reject

Therefore we conclude that there is reasonable evidence that faith and geographical location (are/are not) independent.   are not

15

A thousand individuals are classified according to sex and according to whether they are colour-blind.

	Male	Female	Total
Not colour-blind	442	514	956
Colour-blind	38	6	44
Total	480	520	1000

What conclusions regarding colour-blindness can we draw from these data ?

Null hypothesis: Assume the two criteria
                          (i.e. sex and colour-blindness) are independent.

Calculate expected frequencies.

On the basis of the null hypothesis,

Exp. frequency of males who are not colour-blind = $\dfrac{\ldots \times \ldots}{\ldots}$ $\dfrac{480 \times 956}{1000}$

                                                       = 459 .

The remaining expected frequencies can be calculated by ... .   subtraction

Hence complete the table on p. 28.

Observed and expected frequencies
(The expected frequencies are in brackets)

	Male	Female	Total
Not colour-blind	442(459)	514(...)	956
Colour-blind	38(...)	6(...)	44
Total	480	520	1000

(459)   (497)
(21)   (23)

(Note: Always check that the expected frequencies sum to the correct row and column totals).

Calculate χ^2 using $\chi^2 = \ldots$.

$$\sum \frac{(O-E)^2}{E}$$

Give full working.

$$\chi^2 = \frac{(442-459)^2}{459} + \frac{(514-497)^2}{497} + \frac{(38-21)^2}{21} + \frac{(6-23)^2}{23}$$

$$= (17)^2 \left(\frac{1}{500} + \frac{1}{500} + \frac{1}{20} + \frac{1}{20} \right)$$

$$= 29$$

Determine level of significance using χ^2 distribution with ... d.f. (how many?).

$\underset{\underset{(2-1)}{\uparrow}}{1} \times \underset{\underset{(2-1)}{\uparrow}}{1}$

1 per cent significance level $\chi^2 = \ldots$
5 per cent significance level $\chi^2 = \ldots$

6.64
3.84

The value $\chi^2 = 29$ is obviously significant at the ...... level.
Therefore there is ... ... evidence of a real difference between the observed and expected values and we ... the null hypothesis.
Therefore we conclude that ... .

1%
almost conclusive
reject
there is almost conclusive evidence that sex and colour-blindness are *not* independent.

16.

The four steps in applying the χ^2 test to a contingency table are

STEP 1. Set up ... ...

null hypothesis

STEP 2. Calculate the expected frequencies using *either*

$$\text{exp. freq.} = \frac{\ldots \times \ldots}{\ldots}$$

$$\frac{R \times C}{T}$$

where R = row total, C = column total, T = grand total
or (where possible) by ... .

subtraction

STEP 3.	Calculate a χ^2 value using $\chi^2 = \ldots$	$\dfrac{(O-E)^2}{E}$
STEP 4.	Determine level of significance using χ^2 distribution with ... d.f. where r = no. of rows and c = no. of cols. and hence draw conclusions.	$(r-1) \times (c-1)$

17.

The following table shows the most recent amount of tobacco consumed regularly by women smokers before the onset of their present illness.

	\multicolumn{4}{c}{No. of patients smoking daily}
	1 + cig.	5 + cig.	15 + cig.	Total
Lung carcinoma patients	8	19	15	42
Control patients with diseases other than cancer	12	10	6	28
Total	20	29	21	70

Is there any evidence of a relationship between cancer and smoking?
Give full working.

If you had any difficulties with this problem go to Frame 18.
If not, go to Frame 19.

18.

STEP 1.	Null hypothesis: Assume there (is a/is no) relationship between cancer and smoking.	is no
STEP 2.	Calculate ... ... *for lung carcinoma patients,* the expected number smoking 1 + cig. daily is calculated (using $\dfrac{R \times C}{T}$ by subtraction).	expected frequencies using $\dfrac{R \times C}{T}$
	Therefore expected frequency = $\dfrac{\ldots \times \ldots}{\ldots}$	$\dfrac{20 \times 42}{70}$
	Similarly, the expected number smoking 5 + cig. daily = ...	$\dfrac{29 \times 42}{70}$
	All the remaining expected frequencies (can/cannot) be calculated by subtraction.	can

Hence complete the following table.

Observed and expected frequencies
(expected frequencies are in brackets)

	1 + cig.	5 + cig.	15 + cig.	
Lung carcinoma patients	8(12)	19(17.4)	15(…)	42
Control patients	12(..)	10(…)	6(…)	28
	20	29	21	70

12	17.4	12.6
8	11.6	8.4

STEP 3.   Calculate χ^2 using

$$\chi^2 = \sum \frac{\cdots}{\cdots}$$

$$\sum \frac{(O-E)^2}{E}$$

Therefore $\chi^2 = \ldots$

$$\chi^2 = \frac{(8-12)^2}{12} + \frac{(19-17.4)^2}{17.4}$$

(Give full working).

$$\frac{(15-12.6)^2}{12.6} + \frac{(12-8)^2}{8}$$

$$\frac{(10-11.6)^2}{11.6} + \frac{(6-8.4)^2}{8.4}$$

which gives $\chi^2 = 4.7$

STEP 4.   Determine level of significance using χ^2 distribution with … (How many?) d.f.

$r \searrow 1 \quad c \swarrow 1$
$\quad 1 \times 2 = 2$

1 per cent significance level $\chi^2 = \ldots$
5 per cent significance level $\chi^2 = \ldots$

9.21
5.99

The value $\chi^2 = 4.7$ is not significant at … level.
Thus draw conclusions.

5%
No evidence of a real difference, therefore accept null hypothesis. We conclude that there is no firm evidence from this data of a relationship between smoking and lung cancer.

19.

A random sample of 30 adults are classified according to sex and the number of hours they watch television during a week.

	Male	Female	Total
Over 25 hours	5	9	14
Under 25 hours	9	7	16
Total	14	16	30

Test the hypothesis that a person's sex and time of watching television are independent.
Give full working.

A complete solution to this problem is given below.

Note: The χ^2 test can only be applied if we have *total* numbers of observations, *not* percentages.

Summary

Application of χ^2 test to an r × c contingency table

STEP 1. Set up null hypothesis.

STEP 2. Calculate expected frequencies using *either* expected
frequency $= \dfrac{R \times C}{T}$
where R = row total, C = column total and T = grand total.
or (where possible) by subtraction.

STEP 3. Calculate a χ^2 value using $\chi^2 = \Sigma \dfrac{(O-E)^2}{E}$.

STEP 4. Determine level of significance using χ^2 distribution with $(r-1) \times (c-1)$ degrees of freedom and hence draw conclusions.

Significance at 1 per cent level — almost conclusive evidence of real difference between observed and expected values.
Therefore reject null hypothesis.

Significance at 5 per cent level — reasonable evidence of real difference between observed and expected values.
Therefore reject null hypothesis.

No significance — no evidence between observed and expected values.
Therefore accept null hypothesis.

Solution to problem given in Frame 19

Null hypothesis: Assume that a person's sex and time of watching television are independent

Exp. frequency of males viewing for over 25 hours $= \dfrac{14 \times 14}{30}$

$= 6.5$

Remaining expected frequencies can be calculated by subtraction.

Observed and expected frequencies

	Male	Female	Total
Over 25 hours	5(6.5)	9(7.5)	14
Under 25 hours	9(7.5)	7(8.5)	16
Total	14	16	30

$$\chi^2 = \frac{(5-6.5)^2}{6.5} + \frac{(9-7.5)^2}{7.5} + \frac{(9-7.5)^2}{7.5} + \frac{(7-8\cdot5)^2}{8.5}$$

which gives $\chi^2 = 1.21$.

For χ^2 distribution with 1 d.f.
1 per cent significance level $\chi^2 = 6.64$
5 per cent significance level $\chi^2 = 3.84$.

The value $\chi^2 = 1.21$ is obviously not significant.
Therefore there is no evidence of a real difference between the observed and expected values and we accept the null hypothesis.

Therefore we conclude that there is no evidence that a person's sex and time of watching television are dependent.

Continuity Correction for 2 × 2 contingency tables

The solution to the problem in Frame 19, given above, enables us to make a note of a special method *for 2 × 2 contingency tables only*. The approximation to the χ^2-distribution is improved if the quantity $(O-E)$ is reduced in absolute magnitude by $\frac{1}{2}$ before calculation. For the problem in Frame 19 we would get

$$\chi^2 = \frac{1^2}{6\cdot5} + \frac{1^2}{7\cdot5} + \frac{1^2}{7\cdot5} + \frac{1^2}{8\cdot5}$$

$$= 0\cdot538$$

In this example the value of χ^2 has been changed drastically, because the difference $(O-E)$ was small. However, use of the correction will not often lead to very different conclusions.

Note for Frame 13 (p. 7)

An explanation of the number of degrees of freedom for χ^2 in contingency tables is as follows. If there are r rows and c columns in the table, there are rc cells altogether. But in calculating the expected values we have fixed the marginal totals to be equal to those observed. Since both rows and columns add up to the same grand total, there are $(r+c-1)$ independent marginal totals which have been fixed in this way. The number of degrees of freedom is obtained by subtracting the number of fixed marginal totals from the total number of cells,

$$rc - (r+c-1) = (r-1)(c-1).$$

Exercises for Section 2

PHYSICAL SCIENCES AND ENGINEERING

Goodness of fit

1. The grades in a statistics course were as follows

Grade	A	B	C	D	E
Frequency	14	18	32	20	16

    Test the hypothesis that the distribution of grades is uniform.

2. It is suggested that a recurring mechanical breakdown is more likely to occur towards the end of an 8 hour shift than at the beginning.
    The times of breakdown are available over a considerable period and when formed into a frequency distribution give the following table.

Two hour period	1	2	3	4	Total
No. of breakdowns	5	8	7	12	32

    Apply χ^2 to test the hypothesis that the chance of breakdown is constant throughout the shift.

Contingency Tables

1. Three driving test examiners examine students for their ability to drive. Is there any evidence that the standards required by the examiners vary?

	A	B	C	Total
Pass	20	30	40	90
Fail	40	50	20	110
Total	60	80	60	200

33

2. An investigation into the performance of two machines in a factory manufacturing large numbers of the same product gave the following results —

	No. of articles	
	Defective	Not defective
Machine A	42	708
Machine B	36	864

Apply a statistical test in order to find out whether there is any significant difference in the performance of the two machines as measured by the number of defective articles produced.

BIOLOGICAL SCIENCES

Goodness of Fit

1. The grades in a statistics course were as follows:-

Grade	A	B	C	D	E
Freq.	14	18	32	20	16

Test the hypothesis that the distribution of grades is uniform.

2. Times of death are recorded for 120 cases.

Midnight —	3 a.m. —	6 a.m. —	9 a.m. —
12	17	14	15
noon —	3 p.m. —	6 p.m. —	9 p.m. —
19	14	14	15

Test the hypothesis that death is likely at all times.

Contingency Tables

1. Three driving test examiners examine students for their ability to drive. Is there evidence that the standards required by the examiners vary?

	A	B	C	Total
Pass	20	30	40	90
Fail	40	50	20	110
Total	60	80	60	200

2. Fifty individuals are classified according to eye colour and shade of hair. Can we conclude from the data shown below that for these individuals there is a significant connection between eye colour and hair shade?

		Hair		Total
		Light	Dark	
Eyes	Blue	23	7	30
	Brown	4	16	20
	Total	27	23	50

SOCIAL SCIENCES

Goodness of Fit

1. The grades in a statistics course were as follows

Grade	A	B	C	D	E
Freq.	14	18	32	20	16

Test the hypothesis that the distribution of grades is uniform.

2. The number of mean absent from work on selected days each month was noted.

Jan.	Feb.	Mar.	Apr.	May	June
75	83	74	67	63	46

From this data would you say that the time of year had any effect on the number of absentees?

Contingency Tables

1. Three driving test examiners examine students for their ability to drive. Is there evidence that the standards required by the examiners vary?

	A	B	C	Total
Pass	20	30	40	90
Fail	40	50	20	110
Total	60	80	60	200

2.

		Headmasters and dep. heads	Assistants	Total
Social origin	Non-manual	33	90	123
	Manual	17	160	177
	Total	50	250	300

Is there any evidence that a teacher's professional status is associated with his social origin?

Solutions

Section 1

PHYSICAL SCIENCES AND ENGINEERING

Students t-distribution

1. $n = 5, \overline{X} = 14.1, s = 2.1$

   Since σ is unknown $\sigma_{\bar{x}} = \dfrac{s}{\sqrt{n}} = \dfrac{2.1}{\sqrt{5}} = 0.95$ (to 2 d. pl.).

   The transformation $t = \dfrac{14.1 - \mu}{0.95}$ produces the t-distribution with 4 d.f.

   For this distribution the probability of a value lying between $-2.776 \to +2.776$ is 0.95

   Substitution of these values into the transformation $t = \dfrac{14.1 - \mu}{0.95}$ gives the 95% confidence interval for μ.

   Thus 95% confidence interval for μ is $11.49 \to 16.71$

2. Null hypothesis: Assume samples were drawn from the same population
   i.e. assume there is no real difference between the mean compressive strengths.

   $$n_1 = 5, \quad \overline{X}_1 = 143, \quad s_1 = 8.28$$
   $$n_2 = 5, \quad \overline{X}_2 = 138, \quad s_2 = 11.58$$

   Best estimate for is given by

   $$s = \sqrt{\dfrac{[(n_1 - 1)s_1^2 + (n_2 - 1)s_2^2]}{n_1 + n_2 - 2}}$$

   $$= \sqrt{\dfrac{(4 \times 68.55) + (4 \times 134.1)}{8}}$$

$$= \sqrt{\frac{202.65}{2}} = \sqrt{101.33}$$

$$s = 10.07 \text{ (to 2 d. pl.)}$$

Now transform to t-distribution with $(5 + 5 - 2)$ (i.e. 8) d.f. using the transformation

$$t = \frac{(\bar{X}_1 - \bar{X}_2)}{\sigma_{\bar{x}}} - 0 \quad \text{where} \quad \sigma_{\bar{x}} = s\sqrt{\left(\frac{1}{n_1} + \frac{1}{n_2}\right)}$$

In this example, $\sigma_{\bar{x}} = 10.07 \sqrt{\left(\frac{1}{5} + \frac{1}{5}\right)}$

$$= 6.37.$$

Therefore $t = \dfrac{(143 - 138)}{6.37}$

$$= 0.79 \text{ (to 2 d. pl.)}.$$

For the t-distribution with 8 d.f.

    1% significance level is $t = -3.355$ and $t = +3.355$
    5% significance level is $t = -2.306$ and $t = +2.306$

The value $t = 0.79$ is not significant at 5 per cent level.
Therefore, no evidence of a real difference between $\bar{X}_1$ and $\bar{X}_2$ and so we accept null hypothesis.
Therefore we conclude that there is no evidence of a real difference between the mean compressive strengths.

3. Null hypothesis: Assume there is no difference between sample mean and 80.0
In this example $\mu = 80$, $\bar{X} = 84.0$, $s = 1.2$, $n = 9$
Since σ is unknown,

$$\sigma_{\bar{x}} = \frac{1.2}{\sqrt{9}} = 0.4$$

Now transform to the t-distribution with 8 d.f. using the transformation

$$t = \frac{\bar{X} - \mu}{\sigma_{\bar{x}}}$$

$$= \frac{84.0 - 80}{0.4}$$

$$= \frac{4}{0.4} = 10$$

For t-distribution with 8 d.f.

    1% significance level is $t = -3.355$ and $t = +3.355$
    5% significance level is $t = -2.306$ and $t = +2.306$

The value $t = 10$ is significant at 1% level.
Therefore almost conclusive evidence of a real difference between $\bar{X}$ and μ and so we reject the null hypothesis.

Therefore we conclude that there is almost conclusive evidence that the true percentage of manganese is not equal to 80.0.

BIOLOGICAL SCIENCES

Students *t*-distribution

1. $n = 4, \bar{X} = 14.1, s = 2.1$

Since σ is unknown, $\sigma_{\bar{x}} = \dfrac{s}{\sqrt{n}} = \dfrac{2.1}{\sqrt{4}} = 1.05$

The transformation $t = \dfrac{14.1 - \mu}{1.05}$ produces the *t*-distribution with 3 d.f.

For this distribution the probability of a value lying between -3.182 and $+3.182$ is 0.95.

Substitution of these values into the transformation $t = \dfrac{14.1 - \mu}{1.05}$ gives the confidence interval for μ.

Thus 95 per cent confidence interval for μ is $10.76 \rightarrow 17.44$

2. Null hypothesis: Assume there is no difference in the effectiveness of the drugs.

$$n_1 = 10, \quad \bar{X}_1 = 0.75, \quad s_1 = 1.70$$
$$n_2 = 10, \quad \bar{X}_2 = 2.33, \quad s_2 = 2.10$$

The best estimate for σ is given by

$$s = \sqrt{\dfrac{[(n_1 - 1)s_1^2 + (n_2 - 1)s_2^2]}{n_1 + n_2 - 2}}$$

$$= \sqrt{\dfrac{[(9 \times 2.89) + (9 \times 4.41)]}{18}}$$

$$= \sqrt{\dfrac{7.30}{2}}$$

$$= \sqrt{3.65}$$

$$s = 1.9$$

Now transform to the *t*-distribution with $(10 + 10 - 2)$ d.f. i.e. 18 d.f. using the transformation

$$t = \dfrac{(\bar{X}_1 - \bar{X}_2) - 0}{\sigma_{\bar{x}}} \quad \text{where} \quad \sigma_{\bar{x}} = s\sqrt{\left(\dfrac{1}{n_1} + \dfrac{1}{n_2}\right)}$$

Therefore in this example $\sigma_{\bar{x}} = 1.9 \sqrt{\left(\dfrac{1}{10} + \dfrac{1}{10}\right)}$ therefore $\sigma_{\bar{x}} = 0.86$

Therefore $t = \dfrac{(0.75 - 2.33)}{0.85} = \dfrac{-1.58}{0.85} = -1.84$

For the *t*-distribution with 18 d.f.

1% significance level is $t = -2.878$ and $t = +2.878$
5% significance level is $t = -2.101$ and $t = +2.101$

The value $t = -1.86$ is not significant at 5% level.
i.e. no evidence of a real difference between $\overline{X}_1$ and $\overline{X}_2$ and so accept null hypothesis. Therefore from this data we can conclude that there is no evidence of a real difference in the effectiveness of the drugs.

3. Null hypothesis: Assume mean length of the sample is not different from that in Scotland.
   In this example: $\mu = 10.1$ mm, $\overline{X} = 11.5$,
   $s = 1.4$ $n = 9$

Since σ is unknown, $\sigma_{\bar{x}} = \dfrac{1.4}{\sqrt{9}} = 0.47$

Now transform to the *t*-distribution with 8 d.f. using the transformation

$$t = \frac{\overline{X} - \mu}{\sigma_{\bar{x}}} = \frac{11.5 - 10.1}{0.47} = \frac{1.4}{0.47} \simeq 3$$

For the *t*-distribution with 8 d.f.

1% significance level is $t = -3.355$ and $t = +3.355$
5% significance level is $t = -2.306$ and $t = +2.306$

The value $t = 3$ is significant at 5% level.
i.e. there is reasonable evidence of a real difference between $\overline{X}$ and μ and so we reject the null hypothesis. Therefore we conclude that there is reasonable evidence that the mean length of the sample of animals is different from that in Scotland.

SOCIAL SCIENCES

Students *t*-distribution

1. $n = 4$, $\overline{X} = 15.00$, $s = 1.4$

Since σ unknown, $\sigma_{\bar{x}} = \dfrac{s}{\sqrt{n}} = \dfrac{1.4}{\sqrt{4}} = 0.7$

The transformation $t = \dfrac{15.00 - \mu}{0.7}$ produces the *t*-distribution with 3 d.f.

For this distribution the probability of a value lying between -3.182 and $+3.182$ is 0.95.

Substitution of these values into the transformation $t = \dfrac{15.00 - \mu}{0.7}$ gives the 95 per cent confidence

Thus 95 per cent confidence interval for μ is $12.77 \to 17.23$.

2. *Null Hypothesis:* Assume samples were drawn from same population.
   i.e. assume there is no real difference in mean reaction time.

$$n_1 = 8, \quad \bar{X}_1 = 1.9, \quad s_1 = 0.8$$
$$n_2 = 8, \quad \bar{X}_2 = 2.6, \quad s_2 = 0.9$$

The best estimate for σ is given by

$$s = \sqrt{\frac{[(n_1-1)s_1^2 + (n_2-1)s_2^2]}{n_1 + n_2 - 2}}$$

$$= \sqrt{\left[\frac{(7 \times 0.64) + (7 \times 0.81)}{14}\right]}$$

$$= \sqrt{\frac{1.45}{2}} = \sqrt{0.725} = 0.85$$

$$s = 0.85$$

Now transform to t-distribution with $(8 + 8 - 2)$ d.f. (i.e. 14 d.f.) using the transformation

$$t = \frac{(\bar{X}_1 - \bar{X}_2) - 0}{\sigma_{\bar{x}}} \quad \text{where} \quad \sigma_{\bar{x}} = s\sqrt{\left(\frac{1}{n_1} + \frac{1}{n_2}\right)}$$

In this example, $\sigma_{\bar{x}} = 0.85\sqrt{\left(\frac{1}{8} + \frac{1}{8}\right)}$

$$= 0.425$$

Therefore $t = \dfrac{(1.9 - 2.6)}{0.425} = -1.65$.

For the t-distribution with 14 d.f.

1% significance level is $t = -2.977$ and $t = +2.977$
5% significance level is $t = -2.145$ and $t = +2.145$

The value $t = -1.64$ is not significant at 5% level.

Therefore no evidence of a real difference between $\bar{X}_1$ and $\bar{X}_2$ and so we accept null hypothesis.
∴ we conclude that there is no evidence of a real difference between the mean reaction times for the two stimuli.

3. *Null Hypothesis:* Assume there is no difference between sample mean and 0.5 carat.

In this example, $\mu = 0.5, \bar{X} = 0.62,$
$s = 0.06, n = 5$

Since σ unknown, $\sigma_{\bar{x}} = \dfrac{0.06}{\sqrt{5}} = 0.03$

$$\sigma_{\bar{x}} = 0.03$$

Now transforms to t-distribution with 4 d.f. using the transformation

$$t = \frac{\bar{X} - \mu}{\sigma_{\bar{x}}}$$

$$= \frac{0.62 - 0.50}{0.03}$$

$$= \frac{0.12}{0.03}$$

$$= 4$$

For t-distribution with 4 d.f.

1% significance level is $t = -4.604$ and $t = +4.604$
5% significance level is $t = -2.776$ and $t = +2.776$

The value $t = 4$ is significant at 5% level.
Therefore there is reasonable evidence of a real difference between $\bar{X}$ and μ and so we reject the null hypothesis.

Therefore we conclude that there is reasonable evidence that the average weight of the diamonds produced by the process is significantly different from 0.5 carat.

Section 2

PHYSICAL SCIENCES AND ENGINEERING

Goodness of fit

1.

Grade	A	B	C	D	E	Total
Freq.	14	18	32	20	16	100

Null hypothesis: Assume distribution is uniform.
Therefore table of observed and expected frequencies is as follows:-

						Total
Observed	14	18	32	20	16	100
Expected	20	20	20	20	20	100

$$\chi^2 = \Sigma \frac{(O-E)^2}{E}$$

$$= \frac{(14-20)^2}{20} + \frac{(18-20)^2}{20} + \frac{(32-20)^2}{20} + \frac{(20-20)^2}{20} + \frac{(16-20)^2}{20}$$

$$= \frac{1}{20}(36 + 4 + 144 + 0 + 16)$$

$$= \frac{200}{2} = 10$$

Test using χ^2 distribution with 4 d.f.
   1% significance level $\chi^2 = 13.28$
   5% significance level $\chi^2 = 9.49$
The value $\chi^2 = 10$ is significant at 5% level.

Therefore reasonable evidence of a real difference between observed and expected values and so we reject null hypothesis.
Therefore we conclude that there is reasonable evidence that the distribution of grades is *not* uniform.

2.

2 hour period	1	2	3	4	Total
Number of breakdowns	5	8	7	12	32

Null hypothesis: Assume chance of breakdown is constant throughout shift.
Therefore table of observed and expected values is as follows

Observed	5	8	7	12
Expected	8	8	8	8

$$\chi^2 = \Sigma \frac{(O-E)^2}{E}$$

$$= \frac{(5-8)^2}{8} + \frac{(8-8)^2}{8} + \frac{(7-8)^2}{8} + \frac{(12-8)^2}{8}$$

$$= \frac{1}{8}(9 + 0 + 1 + 16)$$

$$= \frac{26}{8} = 3.25$$

Test using χ^2 distribution with 3 d.f.
 1% significance level $\chi^2 = 11.34$
 5% significance level $\chi^2 = 7.82$
The value $\chi^2 = 3.25$ is not significant at 5% level.
Therefore no evidence of real difference between observed and expected values and so we accept null hypothesis.
Therefore we conclude that there is no evidence that the chance of a breakdown is not constant throughout the shift.

Contingency Tables

1.

	A	B	C	Total
Pass	20	30	40	90
Fail	40	50	20	110
Total	60	80	60	200

Null hypothesis: Assume there is no difference in the standards required by the 3 examiners.
Table of observed and expected values can therefore be constructed.

For examiner A, expected number of passes $= \dfrac{60 \times 90}{200} = 27$

using $\dfrac{R \times C}{T}$

For examiner B, expected number of passes $= 36$
Remaining expected frequencies can be calculated by subtraction.
Observed and expected frequencies (Expected frequency in brackets)

	A	B	C	Total
Pass	20(27)	30(36)	40(27)	90
Fail	40(33)	50(44)	20(33)	110
Total	60	80	60	200

$\chi^2 = \Sigma \dfrac{(O-E)^2}{E}$

$= \dfrac{(20-27)^2}{27} + \dfrac{(30-36)^2}{36} + \dfrac{(40-27)^2}{27} + \dfrac{(40-33)^2}{33} + \dfrac{(50-44)^2}{44} + \dfrac{(20-33)^2}{33}$

$= \dfrac{49}{27} + \dfrac{36}{36} + \dfrac{169}{27} + \dfrac{49}{33} + \dfrac{36}{44} + \dfrac{169}{33}$

$= \dfrac{218}{27} + \dfrac{218}{33} + 1 + \dfrac{9}{11}$

$= 16.5.$

Test using χ^2 distribution with 2×1 d.f.
 1% significance level $\chi^2 = 9.21$
 5% significance level $\chi^2 = 5.99$
The value $\chi^2 = 16.5$ is significant at 1% level.
Therefore almost conclusive evidence of real difference between observed and expected values and so we reject null hypothesis.
Therefore we conclude that there is almost conclusive evidence that the standards required by the examiners do, in fact, vary.

2.

	Defective	Not defective	Total
Machine A	42	708	750
Machine B	36	864	900
Total	78	1572	1650

Null hypothesis: Assume there is no significant difference in performance of the two machines as measured by number of defective articles, i.e. assume the two criteria are independent.
Table of observed and expected frequencies can therefore be constructed.

For machine A, expected number of defectives $= \dfrac{78 \times 750}{1650} = 35.5$

$\dfrac{750}{1650} = \dfrac{15}{33} = \dfrac{5}{11}$

The remaining expected frequencies can be calculated by subtraction
Observed and expected frequencies (expected frequencies in brackets)

	Defective	Not defective	Total
Machine A	42(35.5)	708(714.5)	750
Machine B	36(42.5)	864(857.5)	900
Total	78	1572	1650

$$\chi^2 = \Sigma \dfrac{(O-E)^2}{E}$$

$$= \dfrac{(42-35.5)^2}{35.5} + \dfrac{(708-714.5)^2}{714.5} + \dfrac{(36-42.5)^2}{42.5} + \dfrac{(864-857.5)^2}{857.5}$$

$$= (6.5)^2 \dfrac{1}{35.5} + \dfrac{1}{714.5} + \dfrac{1}{42.5} + \dfrac{1}{857.5}$$

$$= 42.25(0.03 + 0.001 + 0.02 + 0.001)$$

$$= 42.25 \times 0.052$$

$$= 2.20.$$

Test using χ^2 distribution with 1 d.f.
    1% significance level $\chi^2 = 6.64$
    5% significance level $\chi^2 = 3.84$
The value $\chi^2 = 2.20$ is not significant at 5% level.
Therefore no evidence of a real difference between the observed and expected values and so we accept null hypothesis.
Therefore we conclude that there is no evidence of a significant difference in the performance of the two machines as measured by the number of defective articles.

BIOLOGICAL SCIENCES

Goodness of fit

1. See physical sciences and engineering – Goodness of fit, No 1.

2. Midnight – 3 a.m. – 6 a.m. – 9 a.m. –
    12   17   14   15

   noon – 3 p.m. – 6 p.m. – 9 p.m. –
    19   14   14   15

Null hypothesis: Assume that death is likely at all times.
Therefore table of observed and expected values is as follows:

									Total
Observed	12	17	14	15	19	14	14	15	120
Expected	15	15	15	15	15	15	15	15	120

$$\chi^2 = \Sigma \frac{(O-E)^2}{E} = \frac{(12-15)^2}{15} + \frac{(17-15)^2}{15} + \frac{(14-15)^2}{15} + \frac{(15-15)^2}{15} +$$

$$\frac{(19-15)^2}{15} + \frac{(14-15)^2}{15} + \frac{(14-15)^2}{15} + \frac{(15-15)^2}{15}$$

$$= \frac{1}{15}(9 + 4 + 1 + 0 + 16 + 1 + 1 + 0)$$

$$= \frac{1}{15} \times 32 \simeq 2.$$

Test using χ^2 distribution with 7 d.f.
    1% significance level $\chi^2 = 18.48$
    5% significance level $\chi^2 = 14.07$.
The value $\chi^2 = 2$ is not significant at 5% level
i.e. no evidence of a real difference between observed and expected values and so accept null hypothesis.
Therefore we conclude that there is no evidence that death is not likely at all times.

Contingency Tables

1. See physical sciences and engineering – Contingency Tables; No 1.

2.

		Hair		
		Light	Dark	Total
	Blue	23	7	30
Eyes				
	Brown	4	16	20
	Total	27	23	50

Null hypothesis: Assume there is no connection between eye colour and hair shade.
Table of observed and expected values can therefore be calculated.
For individual with blue eyes and light coloured hair expected value

$$= \frac{27 \times 30}{50} = \frac{81}{5} = 16.2$$

Remaining expected frequencies can be calculated by subtraction.

Observed and expected frequencies (expected frequencies in brackets)

		Hair		Total
		Light	Dark	
Eyes	Blue	23(16.2)	7(13.8)	30
	Brown	4(10.8)	16(9.2)	20
	Total	27	23	50

$$\chi^2 = \Sigma \frac{(O-E)^2}{E}$$

$$= \frac{(23-16.2)^2}{16.2} + \frac{(7-13.8)^2}{13.8} + \frac{(4-10.8)^2}{10.8} + \frac{(16-9.2)^2}{9.2}$$

$$= (6.8)^2 \left(\frac{1}{16.2} + \frac{1}{13.8} + \frac{1}{10.8} + \frac{1}{9.2} \right)$$

$$= (6.8)^2 (0.063 + 0.071 + 0.090 + 0.109)$$

$$= 46.24 \times 0.333 = 15.3.$$

Test using χ^2 distribution with 1 d.f.
    1% significance level $\chi^2 = 6.64$
    5% significance level $\chi^2 = 3.84$.
The value $\chi^2 = 15.4$ is significant at 1% level.
i.e. almost conclusive evidence of a real difference between observed and expected values and so reject null hypothesis.
We conclude that there is almost conclusive evidence that there is a connection between eye colour and hair shade.

SOCIAL SCIENCES

Goodness of fit

1. See physical sciences and engineering — Goodness of fit; No: 1.

2.

Jan.	Feb.	Mar.	Apr.	May	June	Total
75	83	74	67	63	46	408

Null Hypothesis: Assume time of year had no effect on the number of absentees.
Therefore table of observed and expected values is as follows:

Obs.	75	83	74	67	63	46	Total
Exp.	68	68	68	68	68	68	408

$$\chi^2 = \Sigma \frac{(O-E)^2}{E} = \frac{(75-68)^2}{68} + \frac{(83-68)^2}{68} + \frac{(74-68)^2}{68} + \frac{(67-68)^2}{68} +$$
$$\frac{(63-68)^2}{68} + \frac{(46-68)^2}{68}$$

$$= \frac{1}{68}(49 + 225 + 36 + 1 + 25 + 484)$$

$$= \frac{820}{68} \simeq 12$$

Test using χ^2 distribution with 5 d.f.
   1% significance level $\chi^2 = 15.09$
   5% significance level $\chi^2 = 11.07$.
The value $\chi^2 = 12$ is significant at 5% level.
Therefore reasonable evidence of real difference between observed and expected values and so we reject null hypothesis.
$\therefore$ we conclude that there is reasonable evidence that the time of year did have an effect on the number of absentees.

Contingency Tables

1. See physical sciences and engineering — Contingency Tables; No: 1.

2.

		Headmasters and deputies	Assistants	Total
Social origin	Non-manual	33	90	123
	Manual	17	160	177
	Total	50	250	300

Null Hypothesis: Assume that teachers' professional status is not associated with his social origin.
Table of observed and expected values can therefore be constructed.
For headmasters and deputy heads with non-manual origin, expected frequency

$$= \frac{50 \times 123}{300} = 20.5.$$

Remaining expected frequencies can be calculated by subtraction.
Observed and expected frequencies (exp. freq. in brackets)

	Heads and deputies	Assistants	Total
Non-manual	33(20.5)	90(102.5)	123
Manual	17(29.5)	160(147.5)	177
Total	50	250	300

$$\chi^2 = \Sigma \frac{(O-E)^2}{E} = \frac{(33-20.5)^2}{20.5} + \frac{(90-102.5)^2}{102.5} + \frac{(17-29.5)^2}{29.5} + \frac{(160-147.5)^2}{147.5}$$

$$= \frac{(12.5)^2}{20.5} + \frac{(12.5)^2}{102.5} + \frac{(12.5)^2}{29.5} + \frac{(12.5)^2}{147.5}$$

$$= (12.5)^2(0.05 + 0.01 + 0.03 + 0.006) \simeq 15.6.$$

Test, using χ^2 distribution with 1 d.f.
   1% significance level $\chi^2 = 6.64$
   5% significance level $\chi^2 = 3.84$.
The value $\chi^2 = 15.6$ is significant at 1% level.
Therefore almost conclusive evidence of a real difference between observed and expected values and so we reject null hypothesis.
We conclude that there is almost conclusive evidence that a teachers professional status is associated with his social origin.

Tables

SQUARES AND SQUARE ROOTS

x	x^2	$\sqrt{x}$	$\sqrt{10x}$	x	x^2	$\sqrt{x}$	$\sqrt{10x}$
1.0	1.00	1.000	3.162	5.5	30.25	2.345	7.416
1.1	1.21	1.049	3.317	5.6	31.36	2.366	7.483
1.2	1.44	1.095	3.464	5.7	32.49	2.387	7.550
1.3	1.69	1.140	3.606	5.8	33.64	2.408	7.616
1.4	1.96	1.183	3.742	5.9	34.81	2.429	7.681
1.5	2.25	1.225	3.873	6.0	36.00	2.449	7.746
1.6	2.56	1.265	4.000	6.1	37.21	2.470	7.810
1.7	2.89	1.304	4.123	6.2	38.44	2.490	7.874
1.8	3.24	1.342	4.243	6.3	39.69	2.510	7.937
1.9	3.61	1.378	4.359	6.4	40.96	2.530	8.000
2.0	4.00	1.414	4.472	6.5	42.25	2.550	8.062
2.1	4.41	1.449	4.583	6.6	43.56	2.569	8.124
2.2	4.84	1.483	4.690	6.7	44.89	2.588	8.185
2.3	5.29	1.517	4.796	6.8	46.24	2.608	8.246
2.4	5.76	1.549	4.899	6.9	47.61	2.627	8.307
2.5	6.25	1.581	5.000	7.0	49.00	2.646	8.367
2.6	6.76	1.612	5.099	7.1	50.41	2.665	8.426
2.7	7.29	1.643	5.196	7.2	51.84	2.683	8.485
2.8	7.84	1.673	5.292	7.3	53.29	2.702	8.544
2.9	8.41	1.703	5.385	7.4	54.76	2.720	8.602
3.0	9.00	1.732	5.477	7.5	56.25	2.739	8.660
3.1	9.61	1.761	5.568	7.6	57.76	2.757	8.718
3.2	10.24	1.789	5.657	7.7	59.29	2.775	8.775
3.3	10.89	1.817	5.745	7.8	60.84	2.793	8.832
3.4	11.56	1.844	5.831	7.9	62.41	2.811	8.888
3.5	12.25	1.871	5.916	8.0	64.00	2.828	8.944
3.6	12.96	1.897	6.000	8.1	65.61	2.846	9.000
3.7	13.69	1.924	6.083	8.2	67.24	2.864	9.055
3.8	14.44	1.949	6.164	8.3	68.89	2.881	9.110
3.9	15.21	1.975	6.245	8.4	70.56	2.898	9.165
4.0	16.00	2.000	6.325	8.5	72.25	2.915	9.220
4.1	16.81	2.025	6.403	8.6	73.96	2.933	9.274
4.2	17.64	2.049	6.481	8.7	75.69	2.950	9.327
4.3	18.49	2.074	6.557	8.8	77.44	2.966	9.381
4.4	19.36	2.098	6.633	8.9	79.21	2.983	9.434
4.5	20.25	2.121	6.708	9.0	81.00	3.000	9.487
4.6	21.16	2.145	6.782	9.1	82.81	3.017	9.539
4.7	22.09	2.168	6.856	9.2	84.64	3.033	9.592
4.8	23.04	2.191	6.928	9.3	86.49	3.050	9.644
4.9	24.01	2.214	7.000	9.4	88.36	3.066	9.695
5.0	25.00	2.236	7.071	9.5	90.25	3.082	9.747
5.1	26.01	2.258	7.141	9.6	92.16	3.098	9.798
5.2	27.04	2.280	7.211	9.7	94.09	3.114	9.849
5.3	28.09	2.302	7.280	9.8	96 04	3.130	9.899
5.4	29.16	2.324	7.348	9.9	98.01	3.146	9.950

STUDENTS T-DISTRIBUTION

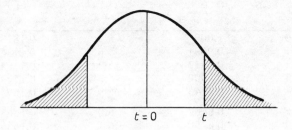

υ or d.f	A = 0.05	A = 0.01
1	12.706	63.657
2	4.303	9.925
3	3.182	5.841
4	2.776	4.604
5	2.571	4.032
6	2.447	3.707
7	2.365	3.499
8	2.306	3.355
9	2.262	3.250
10	2.228	3.169
11	2.201	3.106
12	2.179	3.055
13	2.160	3.012
14	2.145	2.977
15	2.131	2.947
16	2.120	2.921
17	2.110	2.898
18	2.101	2.878
19	2.093	2.861
20	2.086	2.845
21	2.080	2.831
22	2.074	2.819
23	2.069	2.807
24	2.064	2.797
25	2.060	2.787
26	2.056	2.779
27	2.052	2.771
28	2.048	2.763
29	2.045	2.756
30	2.042	2.750
40	2.021	2.704
60	2.000	2.660
120	1.980	2.617
∞	1.960	2.576

RECIPROCALS OF NUMBERS

x	$\frac{1}{x}$	x	$\frac{1}{x}$
1.0	1.000	5.5	0.1818
1.1	0.9091	5.6	0.1786
1.2	0.8333	5.7	0.1754
1.3	0.7692	5.8	0.1724
1.4	0.7143	5.9	0.1695
1.5	0.6667	6.0	0.1667
1.6	0.6250	6.1	0.1639
1.7	0.5882	6.2	0.1613
1.8	0.5556	6.3	0.1587
1.9	0.5263	6.4	0.1562
2.0	0.5000	6.5	0.1538
2.1	0.4762	6.6	0.1515
2.2	0.4545	6.7	0.1493
2.3	0.4348	6.8	0.1471
2.4	0.4167	6.9	0.1449
2.5	0.4000	7.0	0.1429
2.6	0.3846	7.1	0.1408
2.7	0.3704	7.2	0.1389
2.8	0.3571	7.3	0.1370
2.9	0.3448	7.4	0.1351
3.0	0.3333	7.5	0.1333
3.1	0.3226	7.6	0.1316
3.2	0.3125	7.7	0.1299
3.3	0.3030	7.8	0.1282
3.4	0.2941	7.9	0.1266
3.5	0.2857	8.0	0.1250
3.6	0.2778	8.1	0.1235
3.7	0.2703	8.2	0.1220
3.8	0.2632	8.3	0.1205
3.9	0.2564	8.4	0.1190
4.0	0.2500	8.5	0.1176
4.1	0.2439	8.6	0.1163
4.2	0.2381	8.7	0.1149
4.3	0.2326	8.8	0.1136
4.4	0.2273	8.9	0.1124
4.5	0.2222	9.0	0.1111
4.6	0.2174	9.1	0.1099
4.7	0.2128	9.2	0.1087
4.8	0.2083	9.3	0.1075
4.9	0.2041	9.4	0.1064
5.0	0.2000	9.5	0.1053
5.1	0.1961	9.6	0.1042
5.2	0.1923	9.7	0.1031

CHI-SQUARE DISTRIBUTION

A denotes the right tail area for the values of χ^2 given below.
υ denotes the number of degrees of freedom (d.f)

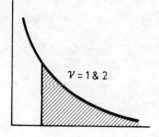
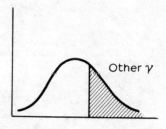

υ or d.f	A = 0.05	A = 0.01
1	3.84	6.64
2	5.99	9.21
3	7.82	11.34
4	9.49	13.28
5	11.07	15.09
6	12.59	16.81
7	14.07	18.48
8	15.51	20.09
9	16.92	21.67
10	18.31	23.21
11	19.68	24.72
12	21.03	26.22
13	22.36	27.69
14	23.68	29.14
15	25.00	30.58
16	26.30	32.00
17	27.59	33.41
18	28.87	34.80
19	30.14	36.19
20	31.41	37.57
21	32.67	38.93
22	33.92	40.29
23	35.17	41.64
24	36.42	42.98
25	37.65	44.31
26	38.88	45.64
27	40.11	46.96
28	41.34	48 28
29	42.56	49.59
30	43.77	50.89